Student Activity Manual

for use with

Technology

R. Thomas Wright

Professor Emeritus, Industry and Technology
Ball State University
Muncie, Indiana

D1501101

Publisher
The Goodheart-Willcox Company, Inc.
Tinley Park, Illinois

INTRODUCTION

You are about to enter an exciting study of technology systems—the technology that makes life better for all of us.

Are you ready to apply your knowledge and skill in challenging new activities? If so, the *Technology Student Activity Manual* is for you. The chapters in this manual will allow you to perform a variety of experiments and build a variety of products. You will not be restricted to a few structured activities. Instead, you can work together with your class and your teacher to select what you will design and produce during your study of technology systems.

As you work through the activities, you will soon realize that there is usually more than one right answer to design and fabrication problems. This manual gives you and your teacher the freedom to select and design the products you will produce. With freedom, however, comes responsibility. You will be responsible for using your creativity and intelligence to contribute to class discussions and decisions. You will also be responsible for maintaining records and notes of class decisions. In addition, you will be responsible for carefully observing demonstrations and recording procedures for completing major processes. Finally, *at all times*, you will be responsible for working safely.

It is wise to set goals. During this class, you will be challenged to accept excellence as your goal. You will be able to select and use technological products more intelligently. You will also become acquainted with many careers in technology.

CONTENTS

	Student Activity Manual	Text
Section 1		
Activity 1A – Design Problem	7	56
Activity 1B – Fabrication Problem	9	56
Section 2		
Activity 2A – Design Problem	13	122
Activity 2B – Fabrication Problem	15	122
Section 3		
Activity 3A – Design Problem	19	174
Activity 3B – Fabrication Problem	23	174
Taking Your Technology Knowledge Home	27	
Section 4		
Activity 4A/4B – Design/Fabrication Problem	33	242
Taking Your Technology Knowledge Home	39	
Section 5		
Activity 5A/5B – Manufacturing Design/Fabrication Problem	45	356
Activity 5C – Construction Design Problem	53	358
Activity 5D – Construction Design Problem	55	358
Activity 5E – Product Servicing Activity	57	
Activity 5F – Papermaking/Paper Recycling Activity	61	
Taking Your Technology Knowledge Home	63	
Section 6		
Activity 6A – Design Problem	69	462
Activity 6B – Production Problem	73	462
Activity 6C – Telecommunication Activity	77	
Taking Your Technology Knowledge Home	79	
Section 7		
Activity 7A – Design Problem	83	522
Activity 7B – Fabrication Problem	85	522
Taking Your Technology Knowledge Home	89	
Section 8		
Activity 8A – Design Problem	95	562
Activity 8B – Fabrication Problem	99	562
Taking Your Technology Knowledge Home	101	
Section 9		
Activity 9A – Design Problem	107	636
Activity 9B – Fabrication Problem	111	636
Taking Your Technology Knowledge Home	113	
Section 10		
Activity 10A – Forming the Company	119	684
Activity 10B – Operating the Company	121	685
Activity 10C – Using and Assessing Activity	131	
Taking Your Technology Knowledge Home	133	
Section 11		
Activity 11A – Design Problem	137	722
Activity 11B – Production Problem	139	722

Section 1

Technology

1 Technology: A Dynamic, Human System

2 Technology As a System

3 Types of Technological Systems

Technology

Class _____

Name _____ Score _____

Activity 1A
Design Problem

Read the **Challenge** for this problem. List the design restrictions contained in the statement.

_____

_____

_____

Sketch two solutions to the design problem on the following grid.

Section 1

List suggestions provided by members of your group for improving the designs.

1. _____   6. _____
2. _____   7. _____
3. _____   8. _____
4. _____   9. _____
5. _____   10. _____

Sketch your final solution to the problem on the following grid.

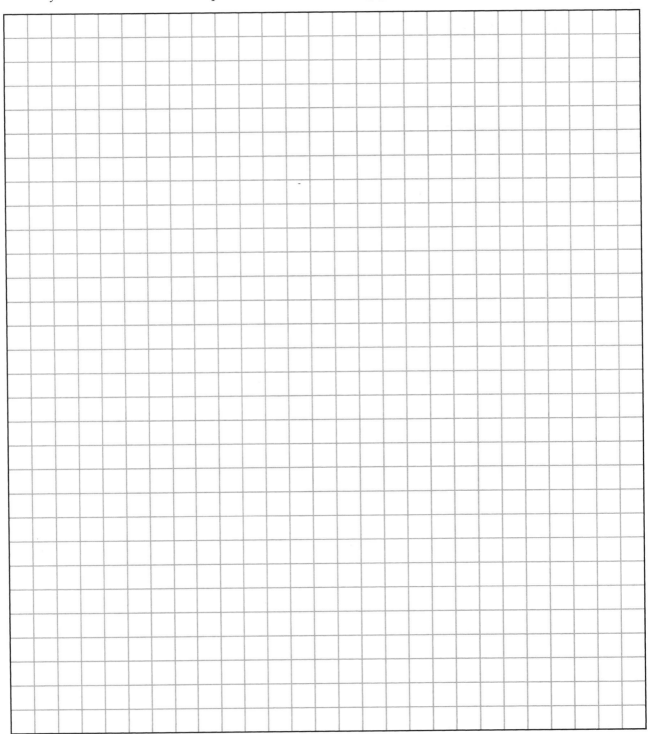

Class _____

Name _____ Score _____

Activity 1B
Fabrication Problem

Read the **Challenge** for this problem. Check the group you are working with:

☐ Group 1: no technology   ☐ Group 2: technology

List the steps you will use to produce the product.

1. _____
2. _____
3. _____
4. _____
5. _____
6. _____
7. _____
8. _____
9. _____
10. _____
11. _____
12. _____
13. _____
14. _____
15. _____
16. _____
17. _____

Section 1

Describe any problems you had in making the product.

Meet as a class and discuss the following questions. Write out your answer for each one.

1. Which production method, with or without technology, represents a more modern manufacturing practice?

2. Which production method, with or without technology, was the easiest to use? Explain your answer.

3. Which production method requires workers with the highest level of skill? Explain your answer.

4. Which production method produces the highest quality outputs? Explain your answer.

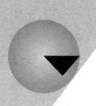

Section 2

Technological System Components

4 Inputs to Technological Systems

5 Technological Processes

6 Outputs and Feedback and Control

Technological System Components

Class _____

Name _____ Score _____

Activity 2A
Design Problem

Read the **Challenge** for this problem. List the materials that you will tentatively include in your "Physics Kit."

1. _____   5. _____   9. _____
2. _____   6. _____   10. _____
3. _____   7. _____   11. _____
4. _____   8. _____   12. _____

On the following grid, sketch two devices that can be built from the list of materials.

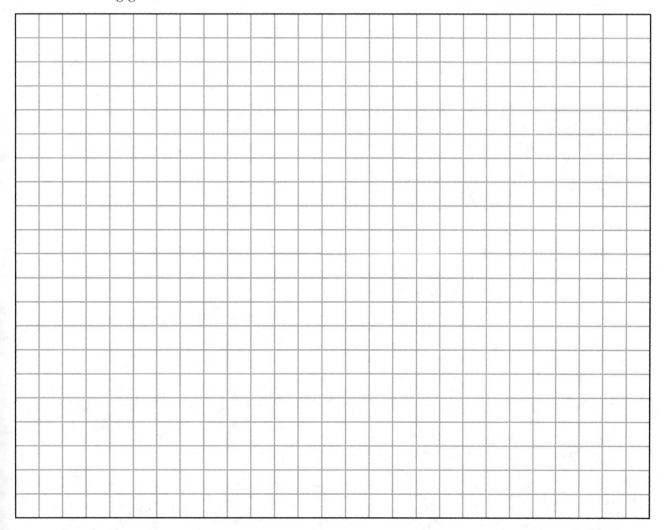

Section 2

List suggestions provided by members of your group for improving the kit and the designs.

1. _____   3. _____   5. _____

2. _____   4. _____   6. _____

In the spaces provided, list the materials that are needed for the improved kit.

1. _____   5. _____   9. _____

2. _____   6. _____   10. _____

3. _____   7. _____   11. _____

4. _____   8. _____   12. _____

On the following grid, sketch a device that can be built from the improved kit.

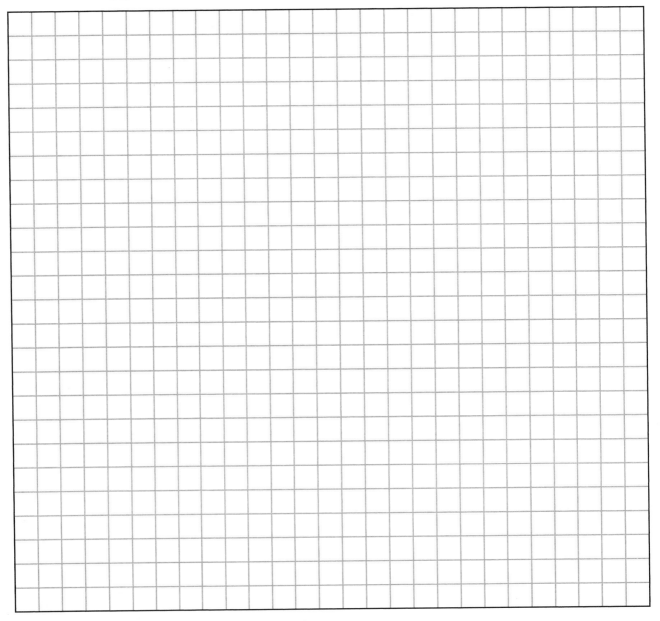

Technological System Components

Class _____

Name _____ Score _____

Activity 2B
Fabrication Problem

List the supplies needed to complete the apparatus shown in the following systems drawing. Label each major part on the line provided.

Systems Drawing

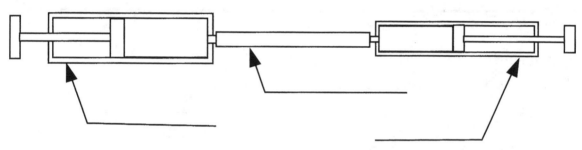

Show how this apparatus can be used as a force multiplier. Draw arrows above each syringe that show their direction of travel. Label one arrow as "FORCE" and the other as "LIFTING."

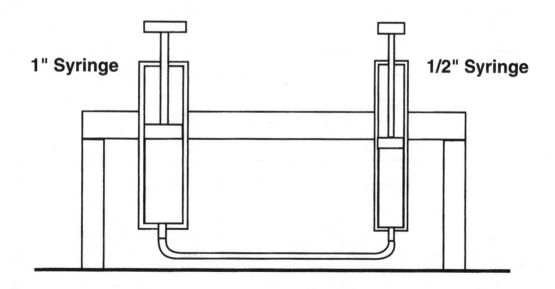

1" Syringe   **1/2" Syringe**

Section 2

Show how this apparatus can be used as a distance multiplier. Draw arrows above each syringe to show their direction of travel. Label one arrow as "FORCE" and the other as "LIFTING."

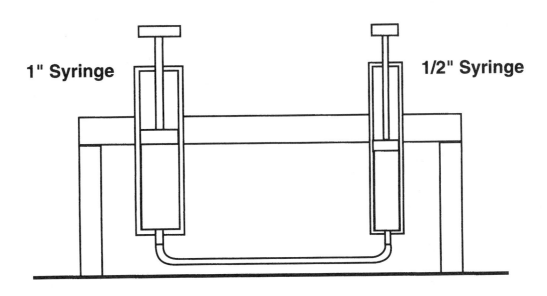

Explain the apparatus used in this activity as a technological system.

Inputs to the system: _____

_____

_____

_____

_____

System processes: _____

_____

_____

_____

System outputs: _____

_____

_____

_____

Section 3

Tools of Technology

7 Production Tools and Their
Safe Use

8 Measurement Systems and Tools
and Their Role in Technology

Tools of Technology

Class _____

Name _____   Score _____

Activity 3A
Design Problem

List three impacts of technology that are not being addressed adequately by the government or private citizens.

1. _____

_____

2. _____

_____

3. _____

_____

Design a flyer to promote citizen awareness of the issue that your group has selected.

Develop the slogan or theme for the flyer._____

_____

Develop the copy (message) for the flyer. _____

_____

_____

_____

_____

_____

_____

_____

_____

Section 3

On the following grid, trace or develop three drawings (line art) that can be used for the flyer.

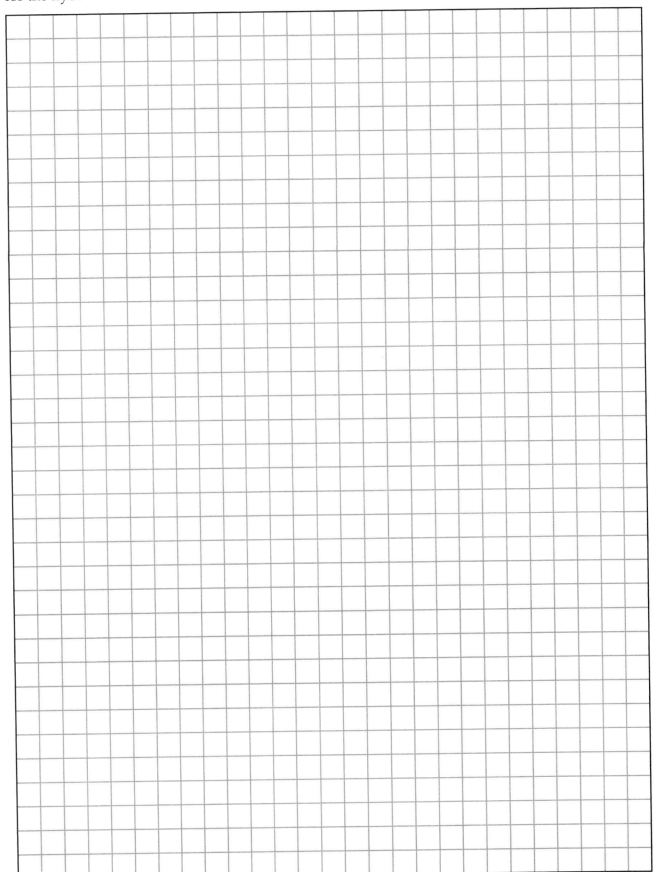

Section 3

On the following grid, develop a layout that integrates your copy and line art into a flyer that will impact people's attitudes about the issue you have selected.

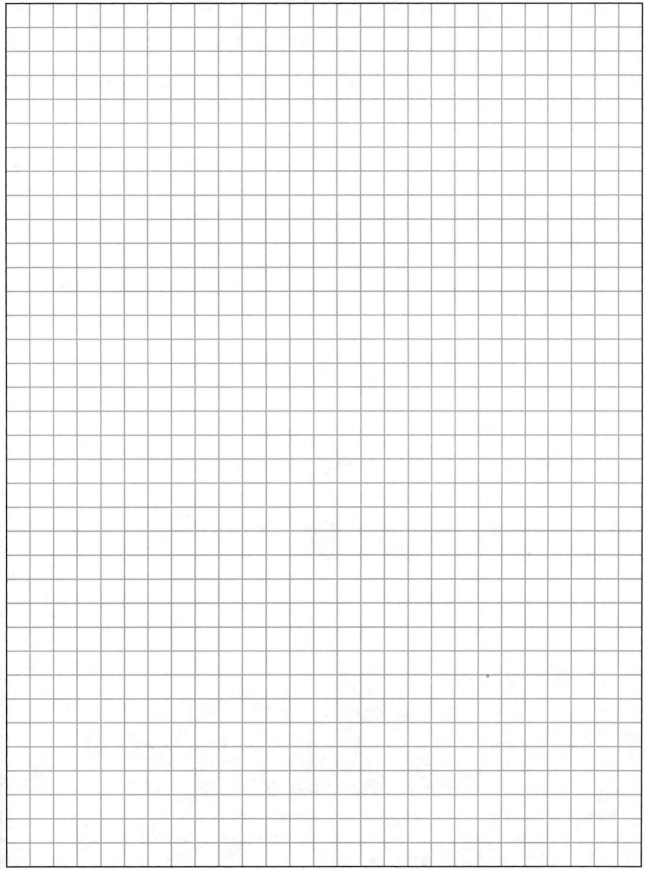

Class _____

Name _____ Score_____

Activity 3B
Fabrication Problem

Have your teacher demonstrate how to dimension a sketch. Then, on the following grid, develop a dimensioned sketch of the product you have selected to build.

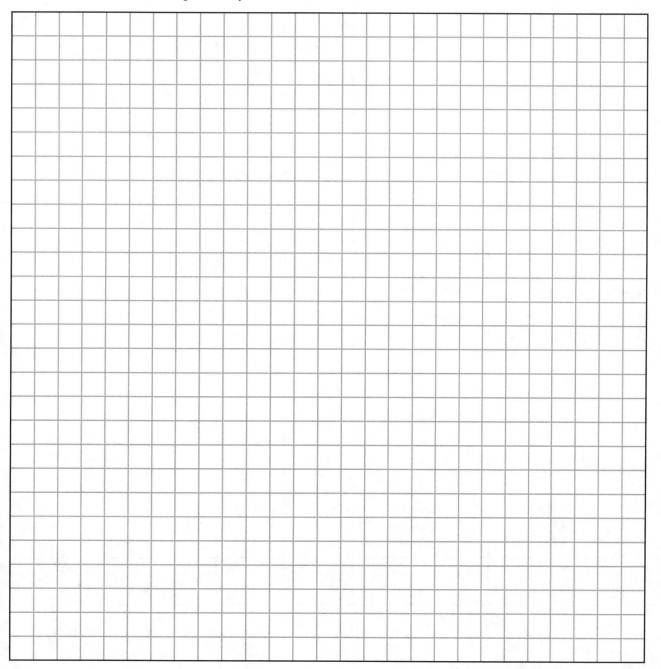

Section 3

As you watch the demonstration of the procedure for making the product, complete the following chart.

Step #	Task	Machine or Tool	Safety Precautions/Considerations										

Section 3

Finish the chart (started on the previous page) on this sheet.

Step #	Task	Machine or Tool	Safety Precautions/Considerations													

Tools of Technology

Class _____

Name _____ Score _____

Taking Your Technology Knowledge Home
Designing a Tool
Introduction

Tools are important to all people. We use tools in almost every task we do. We use writing tools (pencils and pens, for example), measuring tools (rulers and tape measures, for example), kitchen tools (knives, mixing spoons, and measuring cups, for example), and garden tools (trowels and hoes, for example) to name a few. Think of all the tools you use in one day.

Challenge

Design a tool that can be used around your home or in doing yard work.

Step 1: Identify a task for which you could design a tool. _____

_____

Step 2: Look through magazines and catalogs to find tools that do similar jobs.

Step 3: List the feature of the tools you found. _____

_____

_____

_____

_____

Step 4: List the features your tool will have. _____

_____

_____

_____

_____

Section 3

Step 5: On the following grid, sketch four possible designs for the tool you plan to make.

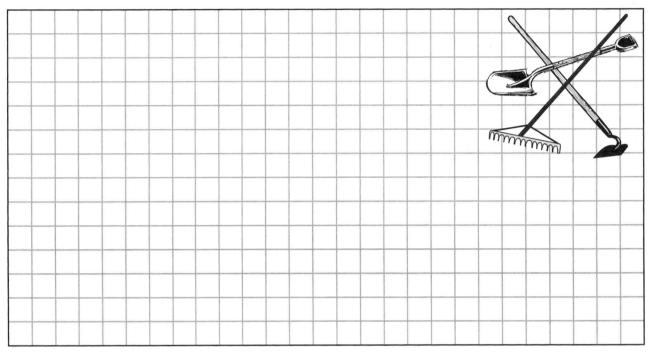

Step 6: Select your best design and, on the following grid, prepare a better (refined) sketch of the tool.

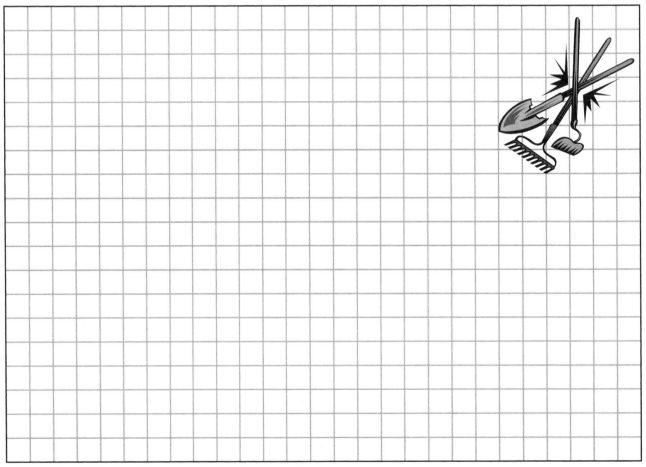

Section 3

Step 7: Build a sample of the tool.

Step 8: Have someone test the tool and comment on its design. Record the comments in the space provided._____

_____

_____

_____

Step 9: On the following grid, sketch a new, improved design for your tool that incorporates the suggestions the users made.

Section 4
Problem Solving and Design in Technology

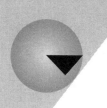

9 The Problem-Solving and Design Process

10 Developing Design Solutions

11 Evaluating Design Solutions

12 Communicating Design Solutions

Problem Solving and Design in Technology

Class _____

Name _____ Score_____

Activity 4A/4B
Design/Fabrication Problem

Write a definition of the design problem before your group.

_____

_____

_____

_____

List points presented by other members of your group that were not stated in your definition.

_____

_____

_____

_____

Write the definition that your group has developed using the several individual definitions presented by its members.

_____

_____

_____

_____

Section 4

List at least three criteria that your product must meet for each category on this sheet.

Engineering criteria

1. _____
2. _____
3. _____
4. _____

Production criteria

1. _____
2. _____
3. _____
4. _____

Marketing criteria

1. _____
2. _____
3. _____
4. _____

Human criteria

1. _____
2. _____
3. _____
4. _____

Financial criteria

1. _____
2. _____
3. _____
4. _____

Environmental criteria

1. _____
2. _____
3. _____
4. _____

Section 4

Develop at least three *rough sketches* of a product that fits your company's product profile. Each sketch should be a complete solution of the product need.

Section 4

Develop a *refined sketch* of the best idea you have for a product that fits your company's product file. Using techniques from Chapters 10 and 12, dimension the sketch.

Section 4

Make a *detailed drawing* of one part of the product your group has chosen. Make sure the design meets the product definition and criteria.

Product name: _____

Part name: _____

Drafter: _____ Class: _____

Scale: _____ Date: _____

Section 4

Produce a *bill of materials* for the product your group has developed.

Bill of Materials

Product: _____

Product Development Group Number: _____

Part Number	Quantity Needed	Part Name	Size			Material
			Thickness	Width	Length	

Problem Solving and Design in Technology

Class _____

Name _____ Score _____

Taking Your Technology Knowledge Home
Designing a Code

Introduction

Codes are used for many things. Manufacturers put codes on products so they know when and where they were made. Most products have bar codes on them so that optical scanners can be used to check inventories and price products at checkout counters.

Codes are also used to send messages. The Morse code was an early example of message codes. It used series of dots and dashes to represent the letters of the alphabet. The codes could be sent as electrical impulses over telegraph wires, as audible sounds over short wave radio, and as flashes of light in ship-to-ship communication.

Language is another code. Sounds we make with our mouth and lips represent ideas and things. For example, the object "desk" is represented by different sounds in different languages.

Code Talkers

One unique use of language codes appeared during World War II. The Navajo language was used as a code throughout the campaign in the Pacific. Code talkers, U.S. servicemen of Navajo descent, transmitted voice messages in their native language. It was a code that the enemy forces never broke.

But why was the Navajo language so effective as a secret code? Its success rested on the fact that it is a very complex, unwritten language. Its sounds and syntax make it difficult to interpret without extensive exposure and training in the language. Also, since the Navajo language is not a written language, it has no alphabet or symbols.

The Navajo code talkers developed a code that used their words to express military terms. A code talker message was a string of seemingly unrelated Navajo words. To translate the message, the code talker translated the Navajo words into English equivalents. (For example, the Navajo word *wol-la-chee* means ant.) The code talker would then take the first letter of each English word to form the decoded word. For example, a message might contain the Navajo words, *tsah, wol-lachee, ah-keh-diglini,* and *tsahah-dzoh.* They would be translated into *needle, ant, victor,* and *yucca.* Taking the first letter of each English equivalent the word navy would appear.

However, to make the code easier to send, some common terms were assigned a specific Navajo word. For example, the Navajo word, *dah-he-tih-hi,* means hummingbird and was used to denote a fighter plane. Likewise the Navajo word, *besh-lo-meas,* means iron fish and was used in place of the word submarine.

The enemy was able to break many codes used in the war, but was never able to break the Marine's Navajo code-talker code.

Section 4
Code Systems

Code systems need a way to hide the meaning in a message (signs or symbols) and a way to send it (technology). The technology involved may use visual devices such as flags, flashes of light, printed symbols, etc., or electrical/electronic equipment such as the telegraph and the radio.

The letter "A" in three different code systems

Morse Code

Sign Language

Semaphore

Challenge

You want a way to communicate to your best friend so that other people will not know the message. Develop a set of symbols or codes that can be used to communicate a message using technology (equipment or devices).

Step 1: Decide how the message will be sent, such as visual images, flashes of light, electronic signals, etc. _____

Section 4

Step 2: Develop a code for each letter and number:

A. ☐ B. ☐ C. ☐ D. ☐ E. ☐

F. ☐ G. ☐ H. ☐ I. ☐ J. ☐

K. ☐ L. ☐ M. ☐ N. ☐ O. ☐

P. ☐ Q. ☐ R. ☐ S. ☐ T. ☐

U. ☐ V. ☐ W. ☐ X. ☐ Y. ☐

Z. ☐ ☐ ☐ ☐ ☐

1. ☐ 2. ☐ 3. ☐ 4. ☐ 5. ☐

6. ☐ 7. ☐ 8. ☐ 9. ☐ 0. ☐

Section 4

Step 3: On the following grid, sketch the technical system (electronic circuit, flag construction, etc.) that you will use for your communication system.

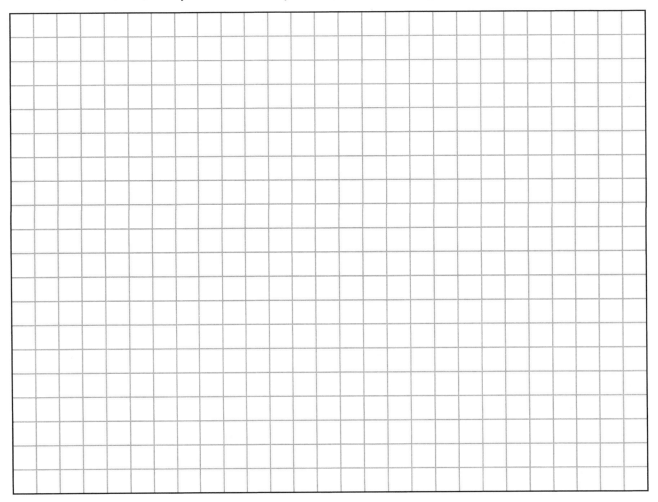

Step 4: Code the phrase "can you read this" in the following boxes.

C

A

N

Y

O

U

R

E

A

D

T

H

I

S

Section 5

Applying Technology: Producing Products and Structures

13 Using Technology to Produce Artifacts

14 The Types of Material Resources and How They Are Obtained

15 Processing Resources

16 Manufacturing Products

17 Constructing Structures

18 Using and Servicing Products and Structures

Applying Technology: Producing Products and Structures

Class _____

Name_____ Score_____

Activity 5A/5B
Manufacturing Design/Fabrication Problem

Sketch the product you will build on the following grid.

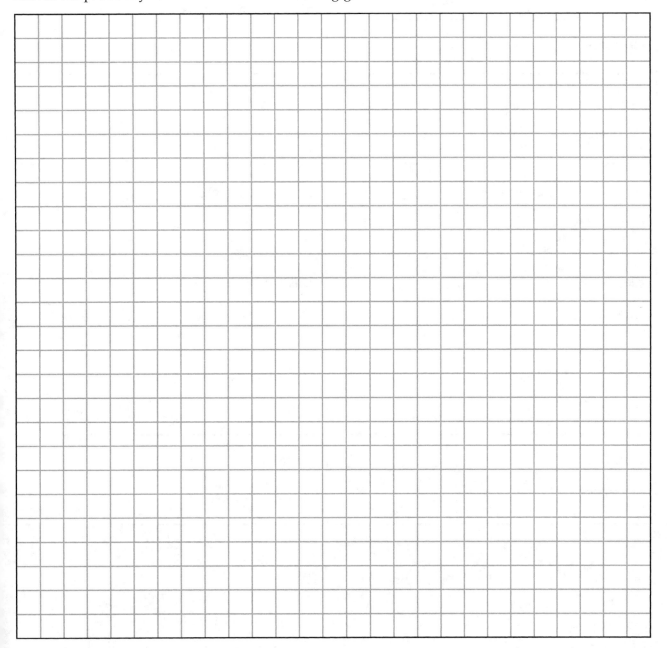

Section 5

Develop a procedure for manufacturing the product on the following form. List each step and the tool or machine that will be used. Identify safety considerations for each process.

Step *sample* Machine: <u>*motor, miter saw*</u> Operation: *Cut the board to length*	Safety considerations: *Keep hand 4 inches from the blade. Check to see guard is functional. Stop blade with brake after each cut.*
Step 1 Machine: _____ Operation:	Safety considerations:
Step 2 Machine: _____ Operation:	Safety considerations:
Step 3 Machine: _____ Operation:	Safety considerations:
Step 4 Machine: _____ Operation:	Safety considerations:
Step 5 Machine: _____ Operation:	Safety considerations:
Step 6 Machine: _____ Operation:	Safety considerations:
Step 7 Machine: _____ Operation:	Safety considerations:

Section 5

Continue the procedure for manufacturing the product on the following form.

Step 8 Machine: _____ Operation:	Safety considerations:
Step 9 Machine: _____ Operation:	Safety considerations:
Step 10 Machine: _____ Operation:	Safety considerations:
Step 11 Machine: _____ Operation:	Safety considerations:
Step 12 Machine: _____ Operation:	Safety considerations:
Step 13 Machine: _____ Operation:	Safety considerations:
Step 14 Machine: _____ Operation:	Safety considerations:
Step 15 Machine: _____ Operation:	Safety considerations:
Step 16 Machine: _____ Operation:	Safety considerations:

Section 5

Select three processes you used. Analyze them on the following form.

Step # _____	Major steps used to complete process:
Description: _____	
_____	
_____	
Process (check one):	
☐ Casting/Molding ☐ Conditioning	
☐ Forming ☐ Assembling	
☐ Separating ☐ Finishing	
Step # _____	Major steps used to complete process:
Description: _____	
_____	
_____	
Process (check one):	
☐ Casting/Molding ☐ Conditioning	
☐ Forming ☐ Assembling	
☐ Separating ☐ Finishing	
Step # _____	Major steps used to complete process:
Description: _____	
_____	
_____	
Process (check one):	
☐ Casting/Molding ☐ Conditioning	
☐ Forming ☐ Assembling	
☐ Separating ☐ Finishing	

Section 5

Select a casting or molding process that you used or observed. Complete the following sheet by describing how each principle was applied.

Casting and Molding			
Mold	☐ Expendable		☐ Cool
	☐ Permanent	Solidify Material	☐ Dry
			☐ Chemical Action
	☐ Melt		
Material Preparation	☐ Dissolve		☐ Open Mold
	☐ Compound	Extract Product	☐ Destroy Mold
Introduce Material	☐ Pour		
	☐ Force		

Major steps used in the process:

Section 5

Select a forming process that you used or observed. Complete the following sheet by describing how each principle was applied.

Forming				
Shaping Device	☐ Die ☐ Roll Type: _____	Method of Applying Force	☐ Press ☐ Hammer ☐ Rolling Machine ☐ Other _____	
Type of Forming Force	☐ Compression ☐ Tension ☐ Bend ☐ Other _____	Material Temperature	☐ Hot ☐ Cold	

Major steps used in the process:

Section 5

Select a separating (machining or shearing) process that you used or observed.
Complete the following sheet by describing how each principle was applied.

Separating		

Cutting
Element   ☐ Tool

            ☐ Other   How the Work Is Supported

Type: _____   _____

Type of   Cutting

Motion   _____

         Feed   How the Work Is Supported

         _____

         _____   _____

Major steps used in the process:

Section 5

Applying Technology: Producing Products and Structures

Class _____

Name_____ Score_____

Activity 5C

Construction Design Problem

Sketch the structure you will build on the following grid.

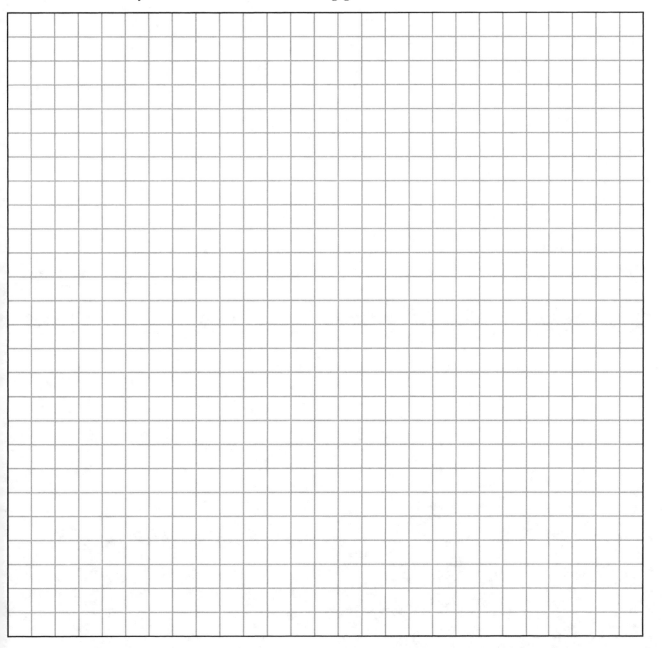

Section 5

Look for design flaws in your structure.

1. _____
2. _____
3. _____
4. _____
5. _____

On the following grid, sketch an improved design for your structure that handles the flaws you listed.

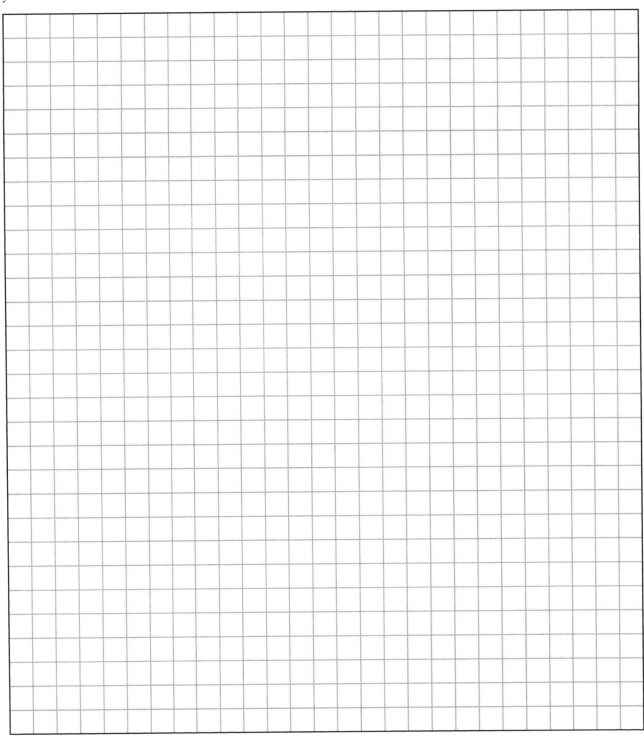

Applying Technology: Producing Products and Structures

Class _____

Name_____ Score_____

Activity 5D
Construction Design Problem

Sketch the structure you will build on the following grid.

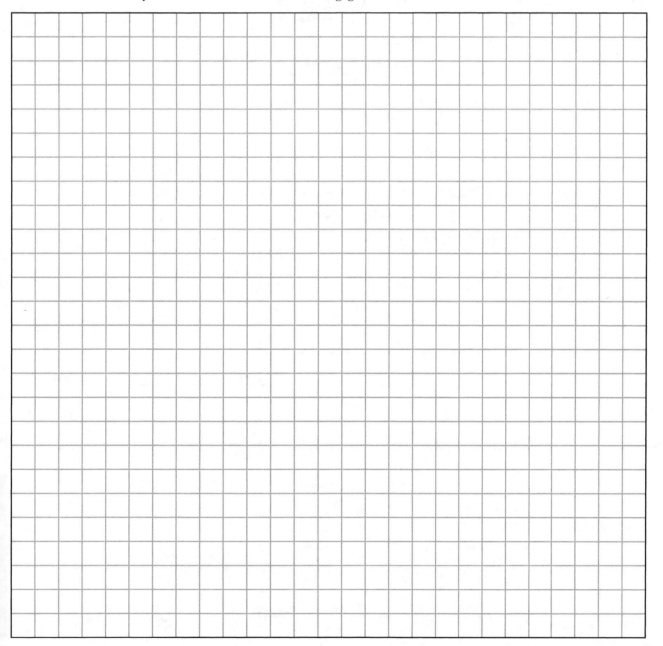

Section 5

Set up the sound insulation experiment as shown in the following illustration.

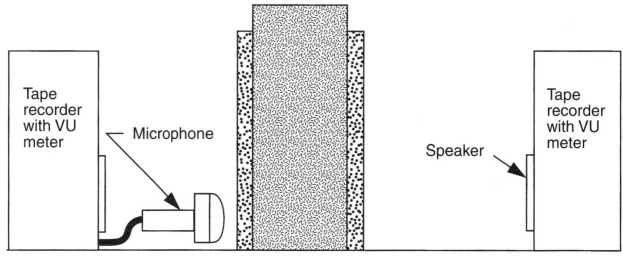

Sound Receiver   Wall Section   Sound Source

Test the sound insulation quality of the wall section using at least five different levels. Use the following steps.

1. Place a recording in the "sound source" tape recorder.
2. Start the tape.
3. Set the volume so that the needle of the VU (volume unit) meter barely moves.
4. Read the VU meter on the "sound receiver" tape recorder.
5. Record the data in the following chart.
6. Repeat steps 1–5 three times. Each time, increase the volume of the source tape recorder.

Test Number	Sound Source Meter Reading	Sound Receiver Meter Reading

Class _____

Name_____ Score_____

Activity 5E
Product Servicing Activity

You will be given a flashlight that will not light. The failure may be a result of probably one (or more) of three areas.

1. The batteries could be dead.
2. The switch could be failing to make contact.
3. The bulb could be burned out.

Service the product by:

1. Reading the maintenance manual.
2. Servicing the product.
3. Completing the service report.

Balsum #601 Flashlight	Service Manual

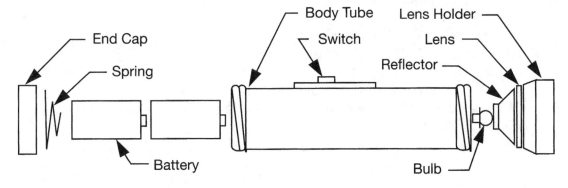

Parts Description

Section 5

The **Balsum #601 Flashlight** seldom needs servicing. However, if it does, the technician should use the following procedure to complete the troubleshooting and repair activities. This procedure follows a logical set of actions that tests the possible areas of failure in the order of their probability, from most probable to least probable.

Testing the Batteries

1. Disassemble the flashlight and remove the batteries.
2. Set the VOM at the 0–10V dc scale.
3. Place the (–) lead of the VOM on the bottom of the battery.
4. Place the (+) lead on the raised tip on the top of the battery.
5. A good battery will produce a reading of 1.5 volts on the VOM.
6. If the batteries are bad, replace them.
7. Reassemble the flashlight and test its operation.

Testing the Bulb

1. Disassemble the flashlight and remove the bulb.
2. Set the VOM to check resistance.
3. Place the (–) lead of the VOM on the contact at the bottom of the bulb.
4. Place the (+) lead of the VOM on the metal side of the bulb.
5. If the VOM reads a maximum (infinite) resistance, the bulb is burned out.
6. If the bulb is burned out, replace it.
7. Reassemble the flashlight and test its operation.

Testing the Switch

1. Disassemble the flashlight.
2. Set the VOM to check resistance.
3. Place the (–) lead of the VOM on one switch contact.
4. Place the (+) lead of the VOM on the other contact.
5. Move the switch control. At one setting, the VOM should read almost no resistance. At the other setting, the VOM should read a maximum (infinite) resistance.
6. If the switch reads a maximum resistance at both settings, clean the switch's contacts.
7. Reassemble the flashlight and test its operation.

Section 5

BALSUM MANUFACTURING	Service Report

Customer's name: _____

Product name: _____ Model number: _____

Date received: _____

What's wrong with the product?

TEST RESULTS

Test	Observations	Corrective action taken
Testing the battery:		
Testing the bulb:		
Testing the switch:		

CUSTOMER BILLING

Time start: _____ Time finished: _____ Time worked: _____

x (labor rate) _____

Labor cost ⟶ [ ]

Parts used:			
Item	Number	Cost	Total
BATTERIES			
BULBS			
		Parts cost	

[ ]

TOTAL BILL
(parts + labor)   [ ]

Section 5
Applying Technology: Producing Products and Structures

Class _____

Name_____ Score_____

Activity 5F
Papermaking/Paper Recycling Activity

Obtain the following materials and supplies.

Papermaking frame, which includes:
- outer frame
- inner frame
- screen wire

16 oz. measuring cup

Hand mixer

8" x 10" baking pan

Paper towels

Hand roller

Single-ply toilet paper/old newspaper

Electric iron

Make a sheet of paper using the following procedure:

1. Shred 60 to 70 sheets of single-ply toilet paper or shred 4 to 8 pages of an old newspaper.
2. Place 1.5 quarts (48 oz.) of water in the baking pan.
3. Mix the shredded paper into the water.
4. Let the mixture stand for 10 minutes.
5. Use a hand mixer to blend the mixture for 5 minutes. (It should look like oatmeal. If newspaper is used, the mixture will be a light grey.)
6. Assemble the production frame:
   a. Place the larger frame part on a table.
   b. Place the screen inside the frame so that it rests on the bent flanges of the frame.
   c. Place the smaller frame inside the larger frame to hold the screen in place.
7. Dip the frame into the pulp mixture. (Note: Tip the screen so it enters the mixture at an angle.)
8. Lift the screen from the mixture.
9. Let it drip for 1 to 2 minutes.
10. Clear the space between the inner and outer frame with your fingers.
11. Remove the screen from the large frame as shown in the diagram on the following page.
12. Remove the inner frame from the screen.
13. Blot the screen/paper mixture from the top with a paper towel.

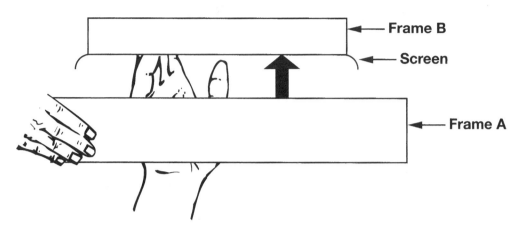

Use the following technique to remove the screen/paper assembly from the frame.

14. Remove the wet paper towel and replace it with a dry one.
15. Carefully roll the screen/paper mixture while the towel is in place.
16. Remove the paper towel, which is now attached to the screen/paper mixture.
17. Place several dry paper towels on a hard surface.
18. Place the towel/paper/screen assembly on the towels with the paper on top.
19. Iron the assembly with an electric iron set at "low" until the paper separates from the assembly.
20. Trim the paper to size.

How would you change the process if you wanted to make 1,000 sheets of paper?

Applying Technology: Producing Products and Structures

Class _____

Name_____   Score_____

Taking Your Technology Knowledge Home
Designing a Manufactured Product
Challenge

Design a product that you and your friends can manufacture for sale or to give to a charity.

Step 1: Identify the use for the product.

☐ Sale

☐ Give to a charity

Step 2: Identify the type of product it will be.

☐ Toy

☐ Game

☐ Household item

☐ Office item

☐ Yard item

☐ Other: _____

Step 3: Specifically define the function of the product. (For example, a device to hold 25 CDs, or a game to entertain children on a trip.)

_____

_____

Step 4: List the feature the product will have.

_____

_____

_____

Step 5: List the limitations you must meet, such as cost, tool available, etc.

_____

_____

_____

Section 5

Step 6: On the following grid, sketch four possible designs for the product you plan to make.

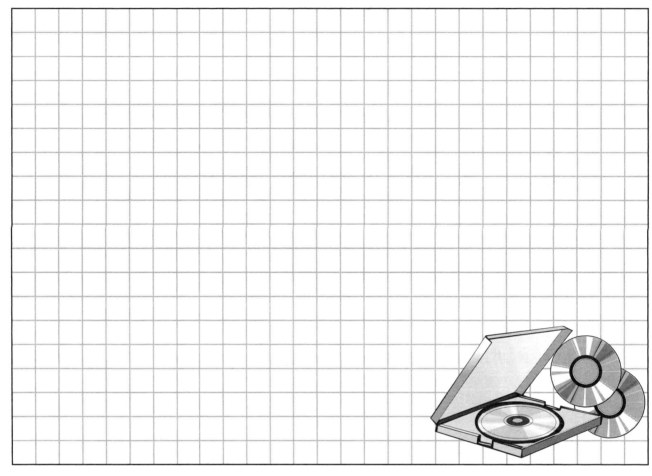

Step 7: Select your best design and, on the following grid, prepare a better (refined) sketch of the product.

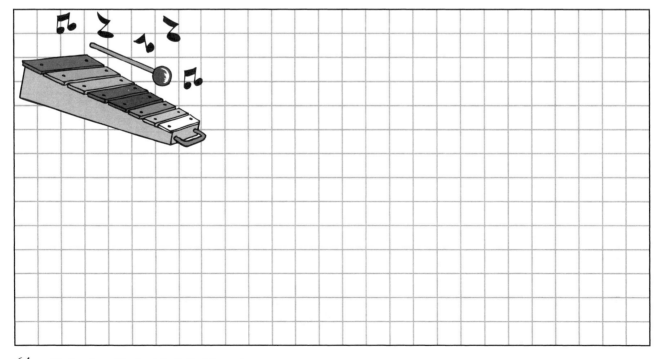

Section 5

Step 8: Build a sample of the product.

Step 9: List the procedure you would use to make ten of the product.

1. _____
2. _____
3. _____
4. _____
5. _____
6. _____
7. _____
8. _____
9. _____
10. _____
11. _____
12. _____
13. _____
14. _____
15. _____
16. _____
17. _____
18. _____
19. _____
20. _____

Step 10: On the following grid, design a package for the product.

Section 6

Applying Technology: Communicating Information and Ideas

19 Using Technology to Communicate

20 Printed Graphic Communication

21 Photographic Communication

22 Telecommunication

23 Computer and Internet Communication

Applying Technology: Communicating Information and Ideas

Class _____

Name_____   Score _____

Activity 6A
Design Problem

Advertising theme:_____

Suggested slogans: _____

Rough sketch #1

Rough sketch #2

Section 6

Rough sketch #3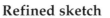

Refined sketch

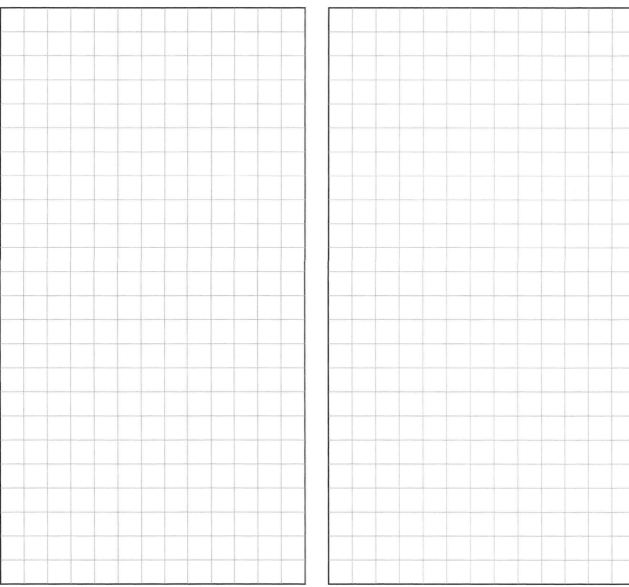

Discussion notes:

Section 6
Comprehensive sketch

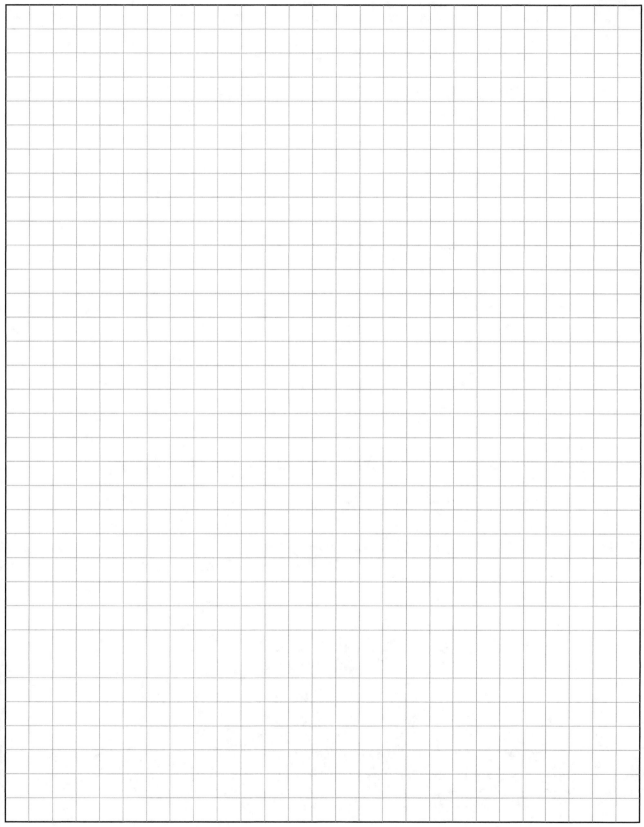

Applying Technology: Communicating Information and Ideas

Class _____

Name_____ Score _____

Activity 6B
Production Problem

Task to be shown in the series of photographs: _____

Steps that are needed to complete the task:

1. _____
2. _____
3. _____
4. _____
5. _____
6. _____
7. _____
8. _____
9. _____
10. _____
11. _____
12. _____
13. _____
14. _____
15. _____
16. _____
17. _____
18. _____
19. _____
20. _____

Six shots that will be used:

1. _____
2. _____
3. _____
4. _____
5. _____
6. _____

Series title: _____

_____

Shot number: _____

Description of the shot:

Series title: _____

_____

Shot number: _____

Description of the shot:

Series title: _____

_____

Shot number: _____

Description of the shot:

Section 6

Series title: _____

_____

Shot number: _____

Description of the shot:

Series title: _____

_____

Shot number: _____

Description of the shot:

Series title: _____

_____

Shot number: _____

Description of the shot:

Applying Technology: Communicating Information and Ideas

Class _____

Name_____ Score _____

Activity 6C

Telecommunication Activity

In this activity you are challenged to create a script for a thirty-second radio or television advertisement. Your instructor will inform you whether the script is for radio or television. (Remember, television scripts will include visual directions and effects; radio scripts will not.)

- Radio
- Television

Your instructor will assign a topic for your advertisement.

Topic: _____

Develop three themes for your topic.

Possible themes: _____

_____

_____

Select the best theme and create a thirty-second script.

SCRIPT		
Character/Event	**Dialog/Directions**	**Production Notes**
Example - - -		
Musical score	*Soft mellow music*	*Fade out to dialog*
Announcer	*Bob and Carol were walking home from school on a windy day*	
Sound effects	*Blowing wind*	*Background growing in intensity*
	Start your script here and complete on next page	

SCRIPT - Page 2		
Character/Event	**Dialog/Directions**	**Production Notes**

Applying Technology: Communicating Information and Ideas

Class _____

Name_____   Score _____

Taking Your Technology Knowledge Home
Promoting a Neighborhood Event

Most neighborhoods have events that interest many people. These events may be garage sales, social gatherings, or charity outings. People need to know about these local happenings. You can help by using your knowledge communication to design and produce a poster or flyer to promote the event. To do this, complete the following steps.

Step 1: Identify the event and sponsor:

- Event: _____
- Sponsor: _____
- Contact person: _____

Step 2: Gather information about the event:

- Date: _____
- Time: _____
- Location: _____
- Cost to attend: _____
- Restrictions: _____
- Purpose: _____
- Other information to be conveyed: _____

  _____

  _____

Step 3: Develop a theme for the poster or flyer. _____

_____

Step 4: Develop a slogan for the poster or flyer. _____

_____

Section 6

Step 5: Develop two possible layouts for the poster or flyer.

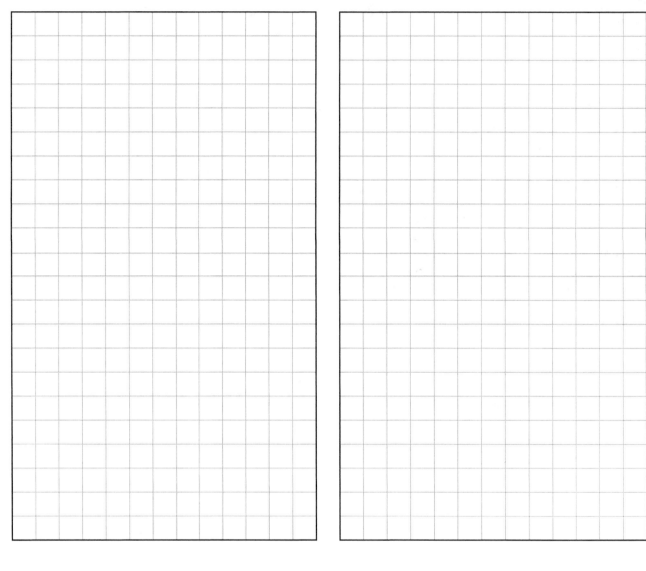

Step 6: Present the layouts to the sponsor and receive suggestions for improvements: _____

_____

_____

Step 7: Produce the poster or flyer.

Applying Technology: Transporting People and Cargo

24 Using Technology to Transport

25 Transportation Vehicles

26 Operating Transportation Systems

Applying Technology: Transporting People and Cargo

Class _____

Name_____ Score _____

Activity 7A
Design Problem

Type of kit: _____

Environment (check one)   ☐ Land   ☐ Water   ☐ Air

Parts List				
Part #	Quantity	Name	Size	Material

Sketch of the Vehicle

Section 7
Assembly Directions

1. _____
2. _____
3. _____
4. _____
5. _____
6. _____
7. _____
8. _____
9. _____
10. _____
11. _____
12. _____
13. _____
14. _____
15. _____
16. _____
17. _____
18. _____
19. _____
20. _____

Section 7

Applying Technology: Transporting People and Cargo

Class _____

Name_____ Score_____

Activity 7B

Fabrication Problem

Vehicle Sketch

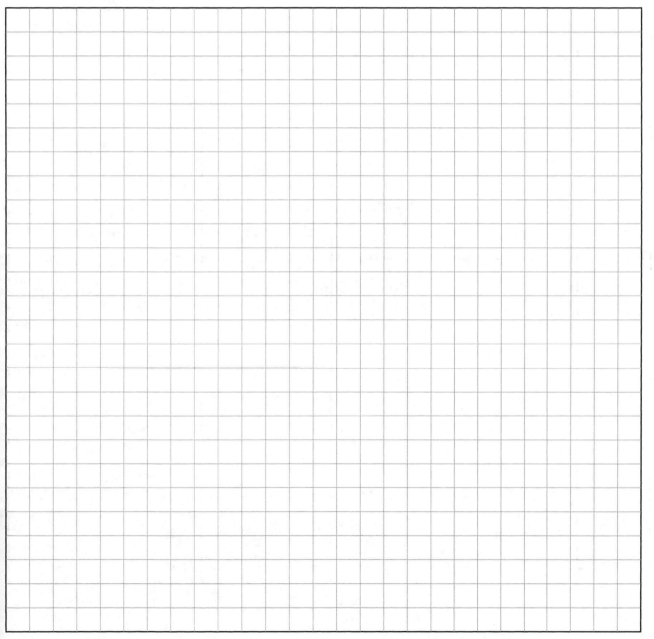

Demonstration Notes			
Step #	Description of Step	Machine/Tool	Safety Considerations

Section 7
Electrical Drawing (Wiring Schematic)

Test Data		

Three-blade propeller:

	Time:	Distance Traveled:
Test #1		
Test #2		
Test #3		

Four-blade propeller:

	Time:	Distance Traveled:
Test #1		
Test #2		
Test #3		

Section 7
Applying Technology: Transporting People and Cargo

Taking Your Technology Knowledge Home
Planning a Trip

Class _____

Name_____ Score_____

Many people use personal transportation systems (family cars) to travel from one place to another. Discuss the possibilities of taking a driving trip with your parents or guardians. Prepare a possible plan for this trip by completing the following steps:

Step 1: Identify the origin (home), destination, and time that can be allowed for the trip. Enter this information in the spaces provided:

Origin: _____

Destination: _____

Length of the trip: _____ days

Step 2: Identify the purpose(s) of the trip using a statement(s) such as "visit my ill grandmother" or "see the Grand Canyon." Write the statement(s) on the lines provided:

Purpose: _____

_____

_____

Step 3: List any special sites or events that can be included on the trip. These can be events such as sporting events, entertainment sites such as theme parks, historical sites such as Civil War battlefields, nature sites such as national parks, or other special places where you want to stop during the trip. On the following lines, list these places or events and the amount of time you will spend at each one.

Site or Event   **Time**

_____   _____

_____   _____

_____   _____

_____   _____

_____   _____

Section 7

Step 4: Use road maps or an Internet mapping site to determine a route for the trip. Consider the places and events you want to visit. Write the driving directions for the trip on the lines below. An example of one entry might be "Leave home and drive 30 miles on Route 34 to Newtown."

_____

_____

_____

_____

_____

_____

_____

_____

_____

_____

_____

_____

_____

_____

_____

_____

_____

_____

_____

_____

_____

Section 7

Step 5: Plan each day's events and costs on a separate 4" x 6" card. Use the following
format to prepare the cards.

Day # _____   Date:_____   Estimated cost for the day: $ _____

Starting point _____   Ending point _____

Miles: _____   Estimated gasoline cost: $ _____

City for: Breakfast: _____   Estimated cost: $ _____

               Lunch: _____   Estimated cost: $ _____

               Dinner: _____   Estimated cost: $ _____

Lodging: _____   Estimated cost: $ _____

Events/Sights:

#1   Estimated cost: $ _____

#2   Estimated cost: $ _____

Step 6: Prepare a summary and a cost estimate for the trip by completing the following chart.

Starting point: _____   Destination: _____

Starting date: _____   Ending date: _____   Round Trip Miles: _____

Estimated costs: Gasoline: $ _____

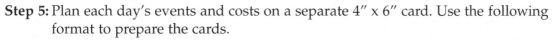

                                          Food: $ _____

                                       Lodging: $ _____

                           Event/Sight-seeing fees: $ _____

                                      TOTAL: $ _____

Section 8
Applying Technology: Using Energy

27 Energy: The Foundation of Technology

28 Energy Conversion Systems

Class _____

Name_____ Score_____

Activity 8A
Design Problem

Check the type of motion your model will display.

☐ Reciprocating   ☐ Pounding

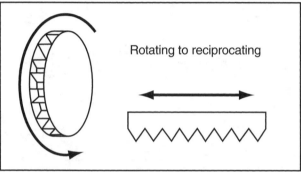
Rotating to reciprocating

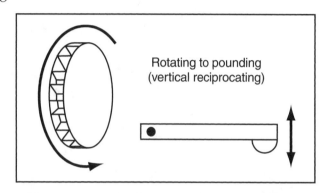
Rotating to pounding (vertical reciprocating)

Prepare rough sketches of three different ways to produce the type of motion your converter is designed to produce.

Rough sketch #1

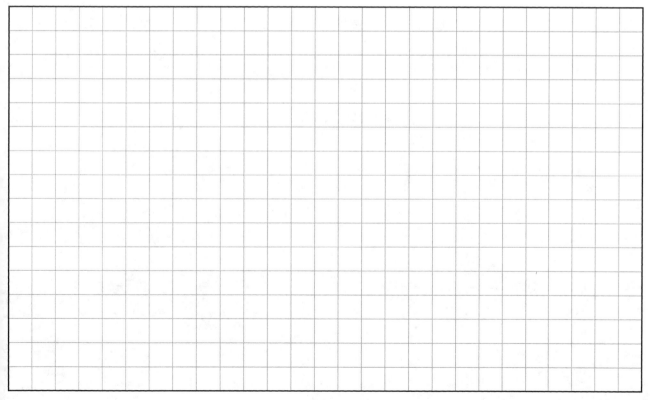

Rough sketch #2

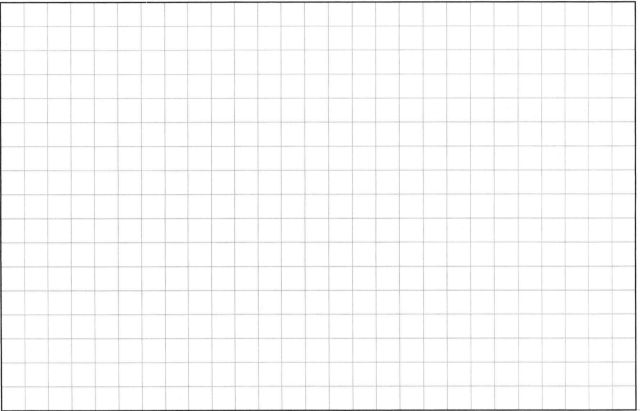

Rough sketch #3

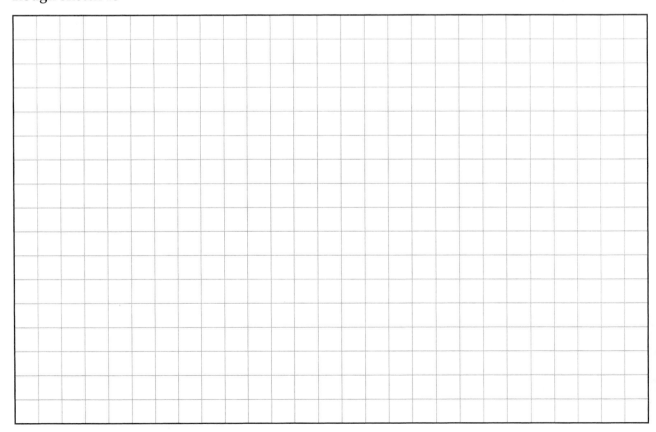

Section 8

Develop a refined, dimensioned sketch of your best design.

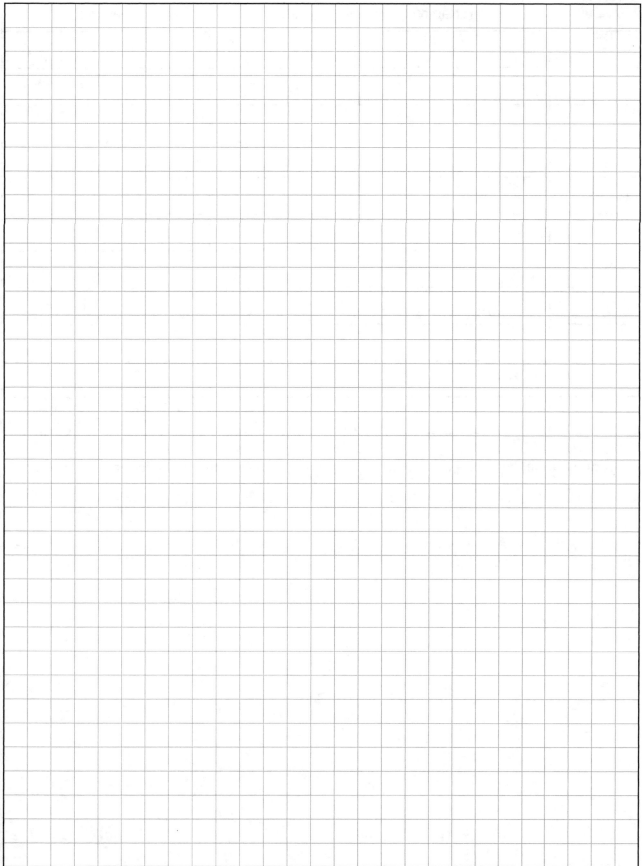

Section 8

Develop a list of materials needed to make the model you have designed.

Quantity	Part Name	Size	Material

Test your model and, in the space provided, list ways you could improve its design.

Applying Technology: Using Energy

Class _____

Name_____ Score_____

Activity 8B
Fabrication Problem

On the following grid, draw a dimensioned sketch of the energy converter you are going to build.

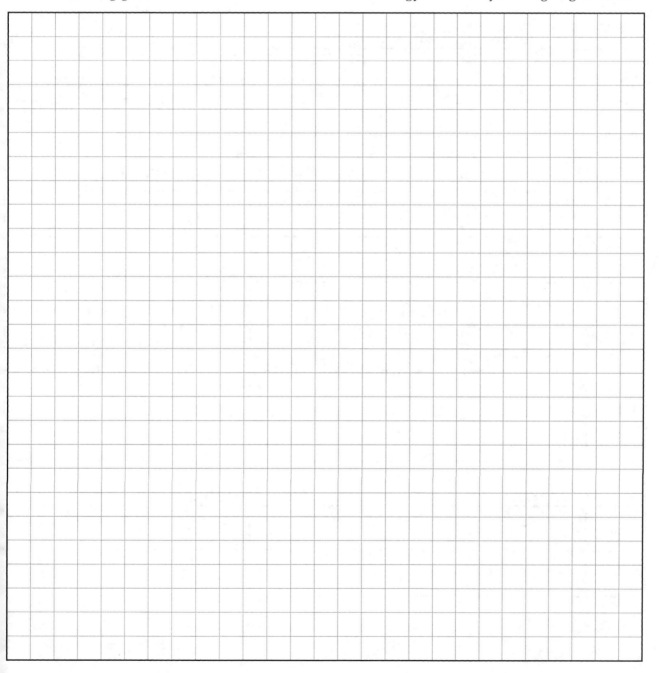

Section 8

Develop a list of materials needed to make your model energy converter.

Quantity	Part Name	Size	Material

List the steps you will use to make the parts for your energy converter.

	Description	Machine	Safety Considerations
1.			
2.			
3.			
4.			
5.			
6.			
7.			
8.			
9.			

Section 8
Applying Technology: Using Energy

Class _____

Name_____ Score_____

Taking Your Technology Knowledge Home
Investigating Lighting Costs and Savings

Energy costs are on the rise. Many homeowners are saving money by replacing standard (incandescent) lighting with compact fluorescent lamps. In this activity, you will survey lighting in at least three rooms of your home and calculate the savings that could be obtained by replacing the bulbs.

Step 1: Locate, identify, and list all the lighting in at least three rooms. Enter the information on the following chart. Make an additional chart on regular paper if you need it.

Present Lighting					
Room	Type of Lamp	Wattage	Hours/Month	Cost/KWH	Cost/Month

Section 8

Step 2: Estimate how many hours each lamp is used per month. Enter this amount on your chart.

Step 3: Determine the kilowatt hours each lamp uses by multiplying the wattage by the hours the light is used. The answer you get is in watt hours of power used in one month. Divide the answer by 1000 to convert the watt hours into kilowatt hours. Enter this amount on your chart.

Step 4: Find out the cost of a kilowatt hour of electricity in your city. Often, this cost is reported on an electric bill. Enter this amount on your chart.

Step 5: Multiply the kilowatt hours used by the cost per kilowatt hour. Enter this amount on your chart.

Step 6: Total the *Cost per Month* column to obtain your present estimated cost of lighting.

Proposed Cost Savings

Now, you can calculate the possible energy and cost savings of replacing all the incandescent lights with compact fluorescent lamps. To do this, complete the following steps:

Step 1: List the rooms and types of lamps on the next chart.

Step 2: Select a proper replacement compact fluorescent for each bulb. Enter the wattage value on the chart.

The recommended replacement bulbs are as follows:

40-watt standard bulb	=	7-watt compact fluorescent bulb
60-watt standard bulb	=	14-watt compact fluorescent bulb
75-watt standard bulb	=	20-watt compact fluorescent bulb
100-watt standard bulb	=	27-watt compact fluorescent bulb
50/100/150-watt 3-way bulb	=	11/22/33-watt compact fluorescent bulb

Step 3: Determine the kilowatt hours each lamp uses by multiplying the wattage times the hours used. The answer you get is in watt hours of power used in one month. Divide the answer by 1000 to convert the watt hours into kilowatt hours. Enter this amount on your chart.

Step 4: Enter the cost of a kilowatt hour of electricity on your chart.

Step 5: Multiply the kilowatt hours used times the cost per kilowatt hour. Enter this amount on your chart.

Step 6: Total the *Cost per Month* column to obtain your present estimated cost of lighting.

Compact Fluorescent Lighting

Room	Type of Lamp	Wattage	Hours/Month	Cost/KWH	Cost/Month

Calculating Savings

Step 1: Subtract the cost of operating the compact fluorescent lighting from the cost of standard lighting. This is the savings per month.

Step 2: Determine the cost of replacing the lamps by using prices from a local store, the Internet, or your teacher.

Step 3: Divide the cost of the lamps by the monthly savings to determine the number of months it will take to pay for the cost of replacing the lamps. Also, compact fluorescent lamps last about five times longer than incandescent bulbs. This longer life also helps pay for the cost of changing bulbs.

Section 9

Applying Technology: Meeting Needs through Biorelated Technologies

29 Agricultural and Related Biotechnologies

30 Food-Processing Technologies

31 Medical and Health Technologies

Applying Technology: Meeting Needs through Biorelated Technologies

Class _____

Name_____ Score_____

Activity 9A
Design Problem

Read the design challenge for this problem. Summarize or state the following:

The problem: _____

_____

Design restrictions or constraints: _____

_____

_____

Sketch two solutions to the design problem on the following grid:

Section 9

Present your design ideas to your group and list the suggestions the members have for improving it.

1. _____
2. _____
3. _____
4. _____
5. _____
6. _____
7. _____
8. _____
9. _____
10. _____

Sketch your final design on the following grid.

Section 9

List the material you will need to build a prototype of your design:

_____   _____

_____   _____

_____   _____

_____   _____

_____   _____

Build and test the prototype. List improvements that should be made.

_____

_____

Sketch an improved design of your device on the following grid.

Applying Technology: Meeting Needs through Biorelated Technologies

Class _____

Name_____ Score_____

Activity 9B
Fabrication Problem

Carefully read the directions for completing the activity.

The problem: _____

Fruit or vegetable selected to preserve by drying:_____

Quantity of produce available: _____

Highest temperature of the dryer: _____

Drying time: _____

Quantity of dried produce: _____

Have several people taste test the dried produce and record their comments:

Tester # 1

Name: _____

Comments:_____

_____

Tester # 2

Name: _____

Comments:_____

_____

Tester # 3

Name: _____

Comments:_____

_____

Applying Technology: Meeting Needs through Biorelated Technologies

Class _____

Name_____ Score_____

Taking Your Technology Knowledge Home
Practicing Primary and Secondary Food Processing

People can practice both primary and secondary food processing at home. Some people grind their own flour, make their own sausage, or can fresh vegetables. Later, they use the output to produce meals. You can practice similar activities at home. To get you started, use this activity to make peanut butter using primary processing techniques and put it on a biscuit you bake using secondary processing techniques.

Primary Food Processing

Primary food processing is the process of changing an agriculture product into a food ingredient.

Peanut butter is a food ingredient that is used in making a number of different products, including snack cracker sandwiches, cookies, and ice cream. To make peanut butter using primary food processing techniques, complete the following steps.

Step 1: Determine the type of peanut butter you will make:

☐ Smooth

☐ Crunchy

Step 2: Obtain the following ingredients:

☐ Roasted peanuts in shells

☐ Corn or vegetable oil

☐ Sugar

Section 9

Step 3: Gather the following processing equipment:

- ☐ Tablespoon
- ☐ Teaspoon
- ☐ Measuring cup
- ☐ Food processor or grinder (for crunchy-style peanut butter)
- ☐ Food blender (for smooth style)
- ☐ Jar

Step 4: Make the peanut butter using the following steps:

1. Open the shells and remove the peanuts.

2. Remove the skin from the peanuts.

3. Measure about one cup of nuts.

4. **For crunchy style:**

   a. Put the peanuts in a food processor or grinder.

   b. Process the mixture three or more times until the peanut butter is the desired consistency.

   For creamy style:

   a. Chop the peanuts in a blender.

5. Add one to two tablespoons of cooking oil slowly while you do Step 6.

6. Bend the mixture until the peanut butter becomes a paste.

7. Add sugar slowly until the desired taste is obtained.

8. Homemade peanut butter contains no preservatives. Therefore, store it in the refrigerator.

Have someone taste your peanut butter and write his or her reactions in the spaces provided.

_____

_____

_____

_____

Secondary Food Processing

Secondary food processing is the process of changing food ingredients into edible products.

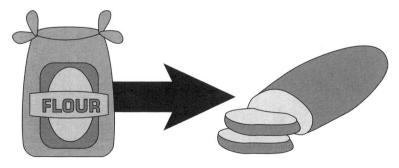

Section 9

Biscuits are bread products that are produced from various food ingredients. To make a batch of biscuits using secondary food processing techniques, complete the following steps.

Step 1: Obtain the following ingredients:

- ☐ 2 cups flour
- ☐ 1/2 teaspoon cream of tartar
- ☐ 1 tablespoon baking powder
- ☐ 1/4 teaspoon salt
- ☐ 2 teaspoons sugar
- ☐ 1/2 cup butter, shortening, or margarine
- ☐ 3/4 cup buttermilk

Step 2: Gather the following processing equipment:

- ☐ Tablespoon
- ☐ Teaspoon
- ☐ Measuring cup
- ☐ Mixing bowl
- ☐ Baking or cookie sheet
- ☐ Jar

Step 3: Make the biscuits using the following steps:

1. Set the oven to 450°F.
2. Sift the flour, baking powder, sugar, cream of tartar, and salt together in a mixing bowl.
3. Cut-in the butter, shortening, or margarine until mixture resembles coarse crumbs.
4. Make a well in the center of the mixture.
5. Add the buttermilk all at once.
6. Stir until dough clings together.
7. Place the dough on a lightly floured surface.
8. Push the dough gently together and flatten to about 1/2 inch thick.
9. Cut discs of dough using a cutter or drinking glass. Dip the cutter in flour between cuts.
10. Transfer the biscuits to a baking sheet.
11. Place the baking sheet in the oven.
12. Bake the biscuits at 450°F for 10–12 minutes or until golden.

Have someone taste your biscuits and write his or her reactions in the spaces provided.

_____

_____

_____

_____

Section 10

Managing a Technological Enterprise

32 Organizing a Technological Enterprise

33 Operating Technological Enterprises

34 Using and Assessing Technology

Managing a Technological Enterprise

Class _____

Name_____   Score_____

Activity 10A
Forming the Company

List the important tasks that are needed to design, produce, and sell a personalized calendar.

1. _____
2. _____
3. _____
4. _____
5. _____
6. _____
7. _____
8. _____
9. _____
10. _____

In the following space, draw an organization chart for a company to design, produce, and market a personalized calendar. The chart should include all departments and show the chain of command (levels of authority and responsibility).

Class _____

Name_____   Score_____

Activity 10B
Operating the Company

Developing the Format

Determine each of the following calendar parameters.

Paper size: _____ Width   _____ Length

Page direction: ☐ Vertical   ☐ Horizontal

Format: ☐ Pictorial   ☐ Full page   ☐ Two month

Day arrangement: ☐ Sunday–Saturday   ☐ Monday–Sunday

Date rectangles: ☐ Square corners   ☐ Rounded corners

Date location: ☐ Upper-left corner   ☐ Upper-right corner

☐ Lower-left corner   ☐ Lower-right corner

Text formats (Fonts):   Text: _____

Credit lines: _____

Month: _____

Day of the week: _____

Date: _____

Credit lines: _____

Pattern in blank date boxes: _____

Other notes:

Promoting Products

On the following grid, develop a layout for a poster promoting your product.

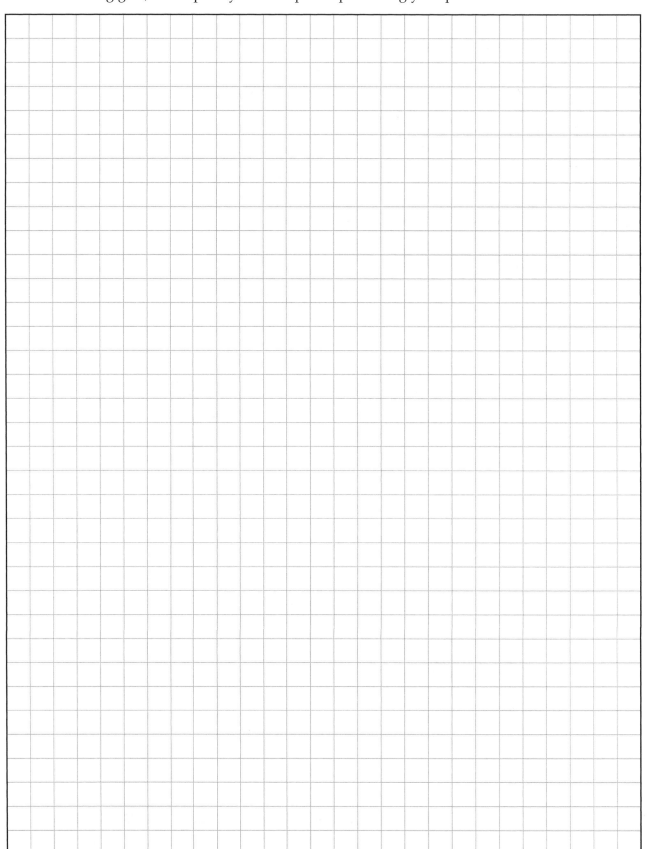

Section 10

Develop a form, or use the following ones, to sell entries on your personalized calendar.

Customer's name: _____

Address: _____ City/State: _____

Telephone number: (_____) _____

Calendar entry:

Date - Month: _____ Day: _____

Credit line entry: _____

Customer's signature: _____

Cost per entry:	$
_____	
Received:	$

Customer's name: _____

Address: _____ City/State: _____

Telephone number: (_____) _____

Calendar entry:

Date - Month: _____ Day: _____

Credit line entry: _____

Customer's signature: _____

Cost per entry:	$
_____	
Received:	$

Customer's name: _____

Address: _____ City/State: _____

Telephone number: (_____) _____

Calendar entry:

Date - Month: _____ Day: _____

Credit line entry: _____

Customer's signature: _____

Cost per entry:	$
_____	
Received:	$

Section 10
Developing Financial Plans

Develop a budget (estimate of income and expenses) for your company.

<table>
<tr><td colspan="3">Income Budget</td></tr>
</table>

Software purchase/rent	$ _____	
Prototype (original model) material	$ _____	
Additional costs	$ _____	
Total development costs	$ _____	(−)

Expense Budget

Production Costs

Image carriers (printing plates)	$ _____	
Paper and other supplies	$ _____	
Labor (hrs) x ($ per hour)	$ _____	
Total production costs	$ _____	(−)

Marketing Costs

Advertising materials	$ _____	
Sales commissions ($ per sale) x (sales)	$ _____	
Total marketing costs	$ _____	(−)

Development Costs

Number of products to sell	_____	
Selling price	$ _____	
Estimated income.	$ _____	(+)
Total Costs	$ _____	

Section 10

Maintain a stockholder record on the following form.

Stockholder's Ledger			
Stockholder's Name	**Address**	**Date Purchased**	**Number of Shares**

Section 10

Develop a form, or use the following ones, to sell your finished personalized calendar.

Customer's name: _____

Address: _____ City/State: _____

Telephone number: (_____) _____

Number of Products	Cost per Product	Total Sale
	$	$

Salesperson _____

Customer's name: _____

Address: _____ City/State: _____

Telephone number: (_____) _____

Number of Products	Cost per Product	Total Sale
	$	$

Salesperson _____

Customer's name: _____

Address: _____ City/State: _____

Telephone number: (_____) _____

Number of Products	Cost per Product	Total Sale
	$	$

Salesperson _____

Section 10

Maintain a record of sales on the following form.

Sales Records											
Salesperson's Name	**Sales by Day**										**Total**
	1	**2**	**3**	**4**	**5**	**6**	**7**	**8**	**9**	**10**	

Section 10
Purchasing Material

Use the following form to inform the Purchasing Department of your material needs.

Purchase Requisition

Recommended vendor:

Company's name: _____

Address: _____

City, State, Zip: _____

Material needed:

Quantity	Description	Est. cost

Authorization:

Charge to budget code number: _____

Approved by Department Head: _____

Approved by Finance Vice President: _____

Section 10

Maintain a record of all financial transactions on the following form.

General Ledger						
Date	**Entry Description**	**Income**		**Expense**	

Managing a Technological Enterprise

Class _____

Name_____   Score_____

Activity 10C
Using and Assessing Activity

Work with two or three other classmates to complete this activity.

Step 1: Many products have owner's manuals that tell people how to use a product. Your product may not need a manual; however, the customer may need some guidance in using it. What information do you think a customer would want so he or she could use the product effectively?

_____

_____

_____

_____

_____

Step 2: Almost all products have impacts on people, society, and the environment. These impacts may be positive or negative. List some impact for your product in the following spaces:

Impact on people
- Positive: _____
- Negative: _____

Impact on society
- Positive: _____
- Negative: _____

Impact on environment
- Positive: _____
- Negative: _____

Managing a Technological Enterprise

Class _____

Name_____   Score_____

Taking Your Technology Knowledge Home
Developing an Organization and Management System

People are constantly organizing events through civic, church, and charitable groups. Each of these events needs to be managed if it is to be run efficiently and effectively. You can help your neighbors and friends by offering to develop a management system for a local event. To do this, complete the following steps:

Step 1: Identify an event for which you could develop a management plan.

- Event: _____
- Sponsor: _____
- Contact person: _____

Step 2: Meet with the contact person to discuss the event and its need for a management structure. Check which of these activities need to be managed.

☐ On-site (production) management (facility setup, security, ticket taking, etc.)

☐ Marketing (promotion, sponsorship, sales, etc.)

☐ Financial affairs (bookkeeping, banking, etc.)

☐ Industrial relations (recruiting and assigning workers, etc.)

Step 3: Identify the major jobs that will have a person(s) assigned to them. Write a brief description for each job. Use 3 × 5 cards like the following one to develop these job descriptions.

> *Ticket Taker:*
>
> *4 Positions*
>
> *Take tickets from attendees. Stamp hands of attendees leaving the event who may wish to return. Check for hand stamps on returning attendees.*

Step 4: On the following grid, develop a management organization chart for the event.

Step 5: Present the job description cards and the organization chart to the event's contact person.

Step 6: Record their comments in the spaces provided.

_____

_____

_____

_____

Section 11

Technological Systems in Modern Society

35 Technology: A Societal View

36 Technology: A Personal View

Technological Systems in Modern Society

Class _____

Name_____   Score _____

Activity 11A
Design Problem

Identify an issue that is important to your community and that is impacted by technology.

The issue: _____

In the spaces provided, list at least five (5) positive and five (5) negative factors related to this use of technology.

Positive Factors

1. _____
2. _____
3. _____
4. _____
5. _____

Negative Factors

1. _____
2. _____
3. _____
4. _____
5. _____

Section 11

Develop a plan of action to address the negative factors of the situation you have identified. List the series of steps that would make up your plan.

1. _____

_____

2. _____

_____

3. _____

_____

4. _____

_____

5. _____

_____

6. _____

_____

7. _____

_____

8. _____

_____

9. _____

_____

10. _____

_____

Section 11

Technological Systems in Modern Society

Activity 11B

Production Problem

Class _____

Name_____ Score_____

Use the following area to develop a storyboard for a TV commercial that promotes recycling.

Series Title: _____

Description:

Description:

Description:

Description:

Section 11

Complete your storyboard on this sheet or any additional sheets you have duplicated.

Description:

Description:

Description:

Description: